James Hasley

Railway Masonry and Bridge Foundations

James Hasley

Railway Masonry and Bridge Foundations

ISBN/EAN: 9783744678728

Printed in Europe, USA, Canada, Australia, Japan

Cover: Foto ©berggeist007 / pixelio.de

More available books at **www.hansebooks.com**

RAILWAY MASONRY

AND

BRIDGE FOUNDATIONS.

BY

JAMES HASLEY,

MASTER MASON LITTLE ROCK & FORT SMITH RAILWAY.

CHICAGO:

THE RAILWAY AGE PUBLISHING CO.

1883.

DEDICATORY.

To THEODORE HARTMAN, of the Little Rock & Fort Smith railway, to whom I owe my sincere thanks for many acts of unostentatious kindness, doubly appreciated when conferred by an employer upon his employe, this little work is respectfully dedicated, with the regret that it is not more worthy,

By the author,
JAMES HASLEY.

ON THE ROAD:
  December 1, 1882.

CONTENTS.

RAILWAY MASONRY.

CHAPTER I.

INTRODUCTION.

I propose in this little work to treat upon the subject of railway masonry, devoting attention more particularly to the foundations and masonry of heavy bridge work, but by no means ignoring those works of lesser magnitude known as culverts. With no impracticable theories to advance, no costly experiments to advocate, I shall on the contrary merely offer such remarks and suggestions as are based upon the actual experience of several years as working mason and foreman in all classes of general railway work in stone, brick and concrete, with the hope that such experience may prove valuable to those who may be interested in the matter.

No one can with propriety question the great superiority of well constructed masonry, over structures of wood, not only in point of solidity and permanence, but also of economy in the long run. Well constructed masonry should never require repairs, never need renewing. Its first cost should be its only cost. Though superstructures should decay and drift away, though embankments should crumble and wash out, masonry should stand as one great mass of solid rock, firm and enduring.

Several items require especial attention in the construction of masonry for water ways, to render it permanent and of value. They are:

1. The selection of stone of proper quality, which will not soften or decompose, crack or scale off from exposure either to air or water.

2. Proper dressing of the stone to true and level beds and reasonable joints.

3. The use of mortars composed of proper materials in proper proportions and thoroughly incorporated with each other.

4. Firm and solid foundations, with the footings at such depth as will cause no danger from the scouring action of the current.

5. Conscientious and thorough workmanship.

Some roadmaster, recently from the scene of a disastrous and expensive washed-out culvert may suggest another item,—that of "Sufficient width of opening." But that in my opinion is entirely a matter of judgment. The stream at its ordinary stage may appear so trivial and insignificant as to require but a narrow opening way. After excessive rainfall the same stream may rise rapidly and swell to considerable proportions, carrying a large volume of water which not finding ample outlet sweeps away the superstructure and washes out the roadbed. But if the masonry is constructed with due regard to the above enumerated items the abutments and piers will remain uninjured and intact. Therefore the disaster could not be attributed to the masonry, as the same would have occurred in the case of a timbered culvert with only equal width of water way, with the great probability that unless well piled and very strongly built the entire wood work, with considerable of the embankment, would have washed out and away.

The width of opening, therefore, is foreign to the subject of masonry, although intimately related to it, and deserving considerable thought, observation and judgment. Ample provision should be made for the passage of unusually large quantities of water, although it might not be required one time in many years.

CHAPTER II.

VARIOUS KINDS OF STONE—STYLES OF MASONRY.

The first point to be observed in the building of masonry is the kind and quality of the stone. For bridge work granite is, beyond all doubt, the most beautiful and durable. Although of several different colors and of varied granulation still it is almost uniformly of the same weight and crushing resistance. This stone being found in large masses and unstratified, in addition to its great hardness, renders it a very costly stone to quarry and dress. I have never seen but one lot of inferior granite, and that was due to the excessive presence of mica.

Most of the limestones are very suitable for pier and abutment work. They occur in layers of various thickness, the stone being very close grained and dressing freely. The blue limestone is very dense and weighs about the same as granite, which is 165 pounds to the cubic foot. The white and magnesian limestones are not so heavy, and when quarried are quite soft, more especially the magnesian, which can be cut with an ordinary pen knife, the same as chalk. Exposure to the air hardens either variety and they answer quite satisfactorily for railway work. Some limestones, however, have a tendency to slake and decompose on exposure to the air and water.

Another very fine stone, which is rarely found imperfect and which preserves its integrity throughout, is the basalt or blue flagstone. It, too, is a stratified stone, with nearly level beds. It splits and works very freely in the direction with its grain, but is exceedingly difficult to work satisfactorily transversely. It is extremely heavy, weighing about 170 pounds to the cubic foot, and can bear much greater transverse strain than any other variety of stone. The smoothness of its beds and its great strength have rendered it the favorite for flagging, curbing and covering of vaults and box culverts. Upon the line of the Little Rock & Fort Smith railway there is a very fine quarry of this stone in layers varying from three inches up to nine and ten. At one point a nine-inch layer is uncovered disclosing one perfect stone 19 feet in length by 7 feet in width, and many others nearly as large.

At Mill Creek the company have a quarry of the same stone existing in heavier layers. I have used many blocks of stone from this quarry measuring seven and eight feet in length by two to three wide, and from two to two and a half in height, weighing from four to five tons each. I have split this stone lengthwise from fifteen to twenty feet with one ounce of powder. Fortunately I was located on the hillside above the track, and simply snubbed it down the hill with a rope and levers onto the cars. At the work I drag it in on skids and rollers operated by a crab and rope. However, this is off the subject, only serving to show at what little cost work can be done with such stone.

Sandstone is the only class which requires the exercise of much judgment and caution. It is so varied in character, its ingredients are so diversified and so dissimilarly incorporated, its texture and weight are so unlike, that considerable inquiry is needed to make sure that it is suitable for railway work. Sandstone is nothing more nor less than sand agglutinated or concreted with soil and mineral substances, such as mica, iron, clay, loam, ochre, marl, etc., etc., in hap-hazard proportions, and its worth as building material is as variable as are the ingredients which compose it.

All sandstone hardens to a greater or less extent upon exposure to the atmosphere, although nearly all that is of large coarse grain and excessive earthly matter will even then be so weak and friable as to be unfit for use on account of the scaling off and crumbling away of the particles.

Sandstone is of various colors and shades (which become very dingy in course of time) and exists in the form of boulders and also in stratas or ledges. It can sustain but little strain across its grain, *i. e.* transversely. In weight it varies from 100 to 130 pounds to the cubic foot and is by far the easiest and cheapest stone to quarry and dress.

The next point to be observed is the proper dressing of the stone to true and level beds and also close joints, if first quality

ORDINARY RUBBLE, WITH DRESSED QUOIN.

work is desired. I have seen some very stable work put up of only the ordinary "rubble," (see illustration) without any dressing except of adraft on the quoins by which to plumb the corners and carry them up neatly. A few strokes of the hammer were also necessary occasionally to spall off any projections or surplus stone. But this style of work is not generally advisable, as very few mechanics could be relied upon to take the proper amount of care in spalling and leveling up the beds and filling the ragged joints of such work. As a consequence one small stone may jar loose and fall out, resulting probably in the downfall of a considerable part of the abutment. However, as most bridge masonry is far more expensive and massive than is generally required, I shall again allude to this class of work under the head of " workmanship."

Another style of cheap work is that known as " broken ashlar" and "bond rubble" or " rubble range work". The broken

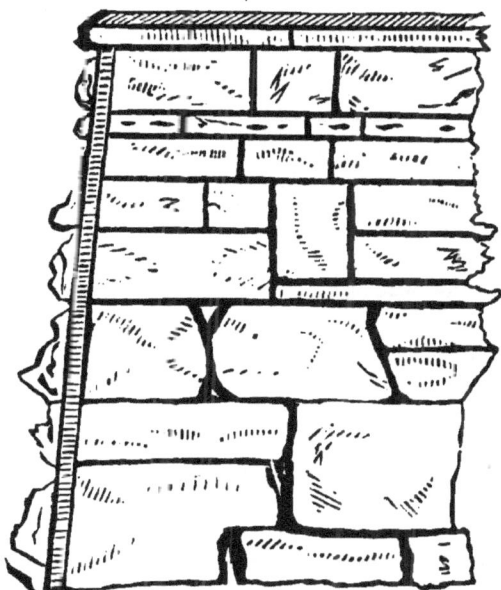

BROKEN ASHLAR, OR RUBBLE RANGE WORK.

ashlar is a dressed stone, but instead of every stone in the course being of the same height, it is admissible to use some of the proper height of the course and also some not of the proper height, but which must be brought up to the level of the others by building more upon them. Bond rubble, or rubble range work is built upon the same style, only of undressed stone, for which the stratified stones are eminently suitable. Any protuberances upon the beds are axed off and a very ragged joint may be roughly hammer-dressed. All of the stratified stones can be laid very economically and to good effect, either in broken or unbroken courses.

Some of these naturally bedded stones are so smooth and uniform as to need no dressing or spalling up. In such case no further search for stone is necessary, and a great expense for cutting is saved and strong massive work is ensured.

CUT ASHLAR, WITH PLAIN FACE AND EDGES.

But the majority of these ready bedded stones have more or less inequalities, and in order to keep them level in the course it becomes necessary to raise them up, perhaps at one corner, perhaps at another or both, as the case may be, by placing a chip or chips of stone, called spalls, under the bed, and slushing the vacant spaces well and full with mortar. And it is just here that the disadvantage of this style of work becomes apparent. Unless the mason distributes these spalls under the stone so that it sets firmly and does not rock,—unless these spalls are placed so that all parts of the stone sustained by them distribute their weight and pressure equally, such stone will in all probability become ruptured and split, and a few such instances in the same work may occasion considerable disaster. But with proper care such class of work under ordinary conditions proves to be all that is needed, and if neatly jointed with good cement mortar will also appear to good advantage.

PITCHED FACED ASHLAR.

The more expensive and generally preferable class of masonry is that known as ashlar. This is cut stone, accurately dressed and laid either in broken or unbroken ranges. Unless the dressing on such work be very precise its superiority ceases to exist. Especially must the bed be cut level and true, so that each part of its surface bears an equal proportion of the entire burden; no hollows, no scantiness, no bumps can be allowed. The aid of spalls should never be needed to set any cut stone in proper position. If it were otherwise what use would there be of expending time and money in cutting such stone? Generally such cutting is done well and accurately, and in works of considerable magnitude ashlar work is most frequently made use of.

CHAPTER III.

In order to please the eye various styles of dressing the faces of stone work have been adopted, and that to be chosen depends upon the taste of the builder. Such dressing is merely for external effect and bears no relation whatever to the strength and durability of the work, for as before remarked, the only points essential are the dressing of the stone to true and level beds and reasonable side joints. Unlike most other work the masonry of railway bridges is very little seen by the general public, to whose eye the builders of other work almost invariably cater. Therefore money expended in costly styles of stone cutting on rail way bridge work serves no practical purpose and gratifies the eye of no one but the builder and the track walker. For these reasons the style of dressing generally chosen for the faces of abutments and piers is the ordinary face of the stone as it leaves the quarry, be that face plain and straight, indented or protruding. This is called quarry or rock faced. If some of the stones have a rather dull, smooth face, and are to be set in a course with others the faces of which are bold and projecting, it is in good taste to pitch the edges of the smooth faced ones off, so as to cause the centre of the face to project to some extent beyond the edges, by which the setting of the course in a straight line is guided. This pitching off produces a more uniform effect, and its cost is so trifling that it would require considerable effort to restrain most builders from adopting pitch faced work.

In this class of work a draft or margin more or less elaborate and distinct must be tooled around the four edges of the face, in order to set the work in line and with the proper batter. These guides may be pitched off quite rapidly with the ordinary pitching tool or wide chisel. There is no necessity whatever for working an accurate margin of precise width and exact tool marks. All that is called for is a sharp, well defined edge, by which to gauge the setting of each stone properly in its course. This style of work, as illustrated in the cut of pitch faced ashlar, is rough, jagged and massive, and in appearance is infinitely more grand and imposing than smooth faced work.

A very beautiful though costly style of work, rarely used except for the corner blocks and tail bonds of the angles of piers and abutments, and then only in such places as to be seen and appreciated, is known as the vermiculated, in which a comparatively smooth faced stone is traversed in all directions by a serpentine net work of deep, sharp tool courses, leaving in the spaces between, warty protuberances of some size, which stand out in bold relief to the tool courses. This style of work requires an accurately tooled margin of equal width and depth drawn around all four of the face edges, in order to set it off properly and as though enclosed in a panel.

A cheaper style of work, which presents a rather elegant effect, is designated the bush hammer dressing. It is used only upon the softer limestones and sandstones, and is produced by a hammer into the face of which has been cut several deep channels, crossing each other at right angles. This leaves the vacant spaces prominent: sharp edged teeth of pyramidal shape. The striking of the hammer on the face of the stone, pits the stone with numberless little indentations, which act as a relief to the otherwise smooth face. This style, too, must be provided with a chiseled margin of exact uniformity. In bridge masonry it is never used, except for the top or covering course, known as the coping. Its use on some stone is objectionable, as it seems to loosen it, and in course of time it is apt to scale off.

The only other style of dressing for railroad work of this character is known as the rubbed face. This is a smooth, plain face, produced by either sawing, ch'seling or rubbing one stone upon the other, as can be seen exemplified at any marble works. Much stone occurs naturally in quarries, with a similar face to that produced by rubbing, and the dry seams and straight grain combine to make it work out very easily and with straight edges and plain face. But such a face is very monotonous in its aspect,—fully as much so as brick. It can, however, be relieved by working a moulding, champer or margin around the edges, and beveling the horizontal joints.

A favorite way of mine in working the blue flag or the limestone, which is always quarried with smooth faces, is simply to relieve the body of the work with sandstone corner blocks and tail bonds, pitch-faced and with drafted joints. It imparts to the work an air of solidity and massiveness which otherwise would be lacking, and the cost of which is trifling.

No matter what the style of the cutting or of the building, the appearance of the work can be greatly enhanced by neat and tasteful pointing of the joints. The mere flushing of the outside joints is all that is absolutely needed as an adjunct to the stability of the work, but if cut stone has been used to produce an effect, the additional expense of pointing up should be incurred. For this purpose a mortar of one part cement, one part lime, and two parts very fine sand is used. Flush the joints full without smearing the faces, then draw straight grooved courses through these mortar joints with the point of a steel setting bar or a tool made expressly for the purpose. Or if the face of the work be smooth, point the joints with the trowel, in the style known as Scotch point, which resembles the puttied bead on the mullion of an ordinary window sash.   On rubble work simply strike the joints by drawing the side of the trowel along them, holding it in an inclined position so as to cut a rather wide V-shaped gash of even depth and width.   But the quickest way to point masonry of any description, is to smooth the mortar joint with the trowel point, which makes an even wide band, and then draw through it a straight, sharp scratch with the edge of the trowel, or, if preferred, draw two parallel scratches.   The neat pointing of any kind of stone work requires considerable time and the exercise of some little skill, and unless the work be of some prominence I make no pretension to pointing up, excepting in very ragged joints which I wish to hide, in which case pointing mortar becomes of as much service as paint and putty to another class of mechanics.

2

CHAPTER IV.

THE FOUNDATIONS.

All necessary details as to the selection of the stone and its proper dressing having been explained, we will advance a step farther, and considering ourselves ready to begin the construction of an open culvert of some importance, will devote ourselves immediately to the consideration of the foundation, bed, bottom or basis upon which we can safely build our masonry with expectation of a very small amount of settlement, and that, in a uniform manner, without rupturing or breaking the bond of the masonry.

It is essential, therefore, that we begin our work upon a firm and stable bottom, one which is practically incompressible, as in the case of bed rock, or one which is nearly so, as in the case of firm level beds of clay, or hard, concreted and indurated gravel.

Every one of my readers has seen great banks of clay, solid, firm, of great cohesion and density, which have served as the walls or confines of some stream for years. All this time the current is gradually but surely cutting out a deep groove or shelf under the bank. No crumbling or falling away of small masses is remarked, but at some hour the last hair's breadth in the groove has been cut (like the last bite of the worm which fells the mighty oak) and the entire bank, in one mass of many tons weight topples over into the stream and lies like a great boulder, without fissure or parting, until completely dissolved. Such a bed of clay is a most admirable foundation for piers and abutments, if of any great extent and depth; it is so impervious to water as to remain solid and never acquire to any great extent that pasty, semi-fluid consistency obtained by ordinary soils and earths.

Every one has also noticed great banks of dark, dingy, brown gravel, coarse grained,.full of iron and concreted firmly together with a natural, earthy cement. This withstands the corroding influence of a rapid current even better than the firm, strong clay just mentioned. I have seen such river banks that seemed to undergo no appreciable change; no marked dissolution or caving was ever noticed; they were the nearest approach in stability, to solid masonry, that I ever witnessed. Solid beds of this, or even of good thick heavy crust overlaying a treacherous substratum, are excellent foundations for bridge work.

Beds of loose, disintegrated gravel and of sand are also nearly incompressible so long as confined and prevented from escaping. But as each particle is separate and distinct from the other, utterly without bond or cohesion, it is powerless to resist the action of running water, and any work built upon such a bottom would be undermined by a rapid current unless far from its reach. In addition to this, much sand is of a quick or mercurial nature; the mere seeping of water into and through it renders it of a semi-fluid character—soft, yielding and treacherous, and consequently such foundations are undesirable upon which to erect bridge masonry. In fact, no matter how firm the ground and river bed upon which a pier is to rest, it must be well protected against any reasonable chance of scour. The depth of the stratum must be considerable, so that the work can be set well down into it, otherwise artificial means must be resorted to, which are generally an apron or toe of heavy rubble or rip-rap stone cast around the pier, so that each stone finds its own bed. In some instances such a procedure would obstruct the channel;

if so the mode of founding must be one yet to be described. Bed rock is the only substance secure against scour and absolutely incompressible, and to novices would seem invariably the most desirable foundation, but in actual practice it is often exactly the reverse. It may present an extremely broken and irregular surface, often smooth and water worn whether level or rounded; in other spots broken into jagged steps and ledges, with numerous cracks and fissures, with a free course for water through them, which renders the labor of leveling the bed quite considerable, the presence of water also making it exceedingly difficult to keep it dry while such leveling is being done.

But if the natural bottom be really rock, not a mere shale or imperfect formation, no compression or settling need be expected except that from the shrinkage of the mortar joints in the courses of the masonry. Our only necessity therefore is to level the foundation pit so that the masonry can start from a level bed. This must be done by blasting and working off the irregularities on the surface of the rock, or by filling up the fissures and vacuities with a reliable and trustworthy concrete, or if perfectly secure from scour, as in the case of a fissure, fine well broken stone or gravel well rammed, will suffice. If, however, the expense be too great to start the work from one level, the excavation may be made with a horizontal step or steps, taking care to execute that part of the work which contains the greatest number of horizontal mortar joints with a harder setting mortar than those parts where the work is not so high; otherwise great difficulty will be experienced in keeping the courses of the stonework level, the settlement being greater according to the increased number or thickness of the horizontal mortar joints.

Generally, at a reasonable depth, such natural bottoms are

found as I have just described. If not, then an artificial bed
must be constructed.

It will therefore be seen that foundations are either natural or
artificial.

To determine upon what manner of foundation our culvert
shall be built, it becomes necessary for us, in the first place, to resort to a preliminary digging, or if at considerable depth, boring
in order to test the bottom. In boring I generally use an ordinary two inch wood auger with jointed rods for lengthening it as
the boring proceeds. This I work with three feet levers, and in
ordinary soils two men will make a boring of twenty or thirty
feet in one day.

But this mode sometimes deceives us in consequence of striking upon a pocket of clay or gravel, or upon a chance boulder,
which as future progress is made in the regular excavation dis-
closes its true character and makes a change of plans unavoida-
ble, as well as an increase of cost.

As a rule, however, we are safe in forming our conclusions
from the indications brought up by the auger. If a natural
foundation is found at reasonable depth we can estimate the cost

ARTIFICIAL FOUNDATION—LOGS FILLED IN WITH CEMENT AND
PLANKED OVER.

of the work. If an artificial foundation is advisable we can decide
as to its kind. In building culverts in which the height and
weight of the masonry are not great, I examine into the nature
of the soil of the banks and bed of the stream, and also ascer-

tain the depth of the bed of the water course. Unless my abutment is to set well back into the bank, perfectly safe from scour and undermining, or is to be protected by a rip-rap toe, I invariably set the foundation several feet below the bed of the stream, varying as I find the tendency of the channel has been to deepen or widen from the scouring of the current.

At what I think to be a sufficient depth for the starting of my footing course, secure from undermining, I cease the work of excavating. If the bottom of the trench be a hard crust of clay or gravel, overlaying soft earth, or if it is an ordinary loamy soil, not very firm, or peaty and full of vegetable fibre as in marshes or in any soil which is somewhat soft and compressible, and not firm enough to bear the weight of the masonry and its load without some equal distribution of the pressure over considerable surface, I level the bottom, and then lay down timbers or sleepers transversely with the excavation, generally using oak ties. These I place eighteen inches apart, filling the vacant places flush with very strong concrete, made of fine broken stone, mixed with equal parts of sand and cement. Upon this bed, longitudinally with the excavation, is laid and spiked a two-inch plank floor, sometimes adding another thickness, which is laid so as to cross the joints of the lower course diagonally, also spiking this layer. This mode makes a very strong bed upon which to start the masonry, and any settlement upon a spongy and compressible soil will be uniform. I have rarely found it necessary to draw back the footing course of stone more than six inches from the edge of the plank flooring, although the greater the extent of flooring the less is the liability of settling.

If the timber is good and the ground wet, which latter is always the case, no fears of decay need be entertained. Under ordinary circumstances, for abutments of culverts this is not only the most inexpensive artificial foundation but is also a reliable one.

The other artificial foundations will receive due attention in future chapters, the next in course being piling and the circumstances under which it is advisable.

It often occurs, however, that the ground is too spongy and compressible to render the artificial foundation just described, advisable. The depth of the soil above the hard stratum may be so great as to make the bringing up of the solid masonry from it an expensive matter, in which case I proceed as follows:

I make the excavation a reasonable depth below low water line, say four feet. I then pile the pit, setting the piles from two and one quarter to five feet apart, from center to center, according to the weight of the masonry to be put upon them. The heads of the piles are cut off level some three or four inches beneath low water line; they are then all connected together by wales or wooden strips, set into shoulders or daps, and spiked.

This then leaves a vacant space between each four piles some three feet and over in depth. This I fill flush up to the top of the heads with concrete of four or five parts finely broken stone mixed up with a mortar of one part cement and two parts sand. I never use a poor concrete. My object is not to see how much sand I can use or how little cement or lime; my purpose is to form an artificial stone which adjusts itself to all the knotty excrescences and unequal shapes of the piles, moulds itself around them tightly and closely and in course of time, if well rammed down, hardens into stone as strong and durable as limestone; the piles then really being imbedded into one solid mass of rock, where if movement or settling takes place, all must be together and uniform.

This is, if properly made, an admirable artificial foundation.

Culvert work never requires any better. The illustration shows this method as adapted to an under water foundation for pier.

PILING FOUNDATION FILLED IN WITH CONCRETE.

Upon the surface of the foundation just described I step with satisfaction and start the footing course of my masonry.

Before commencing my excavation I give some thought to its size and shape. These depend upon, first, the amount of protection, if any, needed for the sides of the embankment. Second, thickness of wall requisite to withstand the outward thrust and pressure of the embankment head.

If the embankment is high the sides will need to be retained by a wing from the abutment proper. A five foot wing extending at an angle of forty-five degrees from the main wall is my favorite, but the stone requires considerable hammer work to shape it for the splayed corners necessary—whereas stone for rectangular corners is much more easily put in shape, for which rea-

son I generally build rectangular wings. If the stream washes
at the foot of the embankment I lengthen the wing at tnat place
accordingly. The higher the embankment the greater the outward pressure. I usually undertake to secure a width of two
and one-half feet at the finish of my work, upon which the wall
plate is laid. In heavy bridge work this same finish at coping
should be five to eight feet, but this is sufficient for culverts.

As I do not wish to have the base of my abutment more than
four or five feet wide, I vary the batter according to its height,
setting back my work as much as one to six for low wall, which
is two inches to one foot in rise. I am now working an abutment twenty feet in height, in which I batter my work one to
twenty-four, or only one inch in two feet rise. I rarely build
any work entirely plumb, because such width of wall is not
needed at the finish, but more especially beca..se a battered wall
preserves itself better against the pressure of the embankment
and is not nearly so easily thrust out of plumb, thereby rendering the falling of the work imminent and disastrous to life
as well as property.

All this decided, I am ready to lay out my pit. Unless an
oblique work is called for I lay out the face of my abutment
proper at right angles with the track. I have had this done very
accurately and to the exact hair's breadth, and amply verified by
. the engineers provided by the company, but as it generally con-
sumes two day's time before these gentlemen are sufficiently sat-
isfied that all stakes are correct and the crosses on the stake
heads in exact position, I weary of the delay and now lay one
off in less than half an hour, which, although perhaps not so
exact as if done with the aid of instruments, is certainly expedi-
tious and sufficiently accurate for all practical purposes.

The following cut shows a method of protecting piling with
a wrought or cast iron shoe, for driving in hard strata.

PILING WITH IRON SHOE.

On the road bed about eight feet back from the face of my
proposed abutment, I establish my center.   I drive a nail in the
tie and with one end of a ten foot pole pressed against the nail I
rest the other end upon one of the rails and chalk the spot.   I
then sweep the pole across to the other rail and chalk it.   I have
simply described the arc of a circle, and a line drawn from the
chalk mark on one rail to the same on the other will be at right
angles to the direction of the roadway, and so if the line is carried on, which I do by a pole from which I drop down my plumb
bob, and drive the stake.   This is done on the other side, and a
line drawn from stake to stake is rectangular with the stringers
above.   If the track is oblique I draw lines from the outside of
one rail at one end of the opening and extend it to the outside of
the rail at the other end.   On the other side I draw another line,
only using the inside of the rails instead of the outside.   Now,

instead of resting one end of my pole upon the actual rails, I mark these strings and obtain thereby a line oblique with the track and in exact degree required. There are other ways of accomplishing the same end, but this suits my purpose best.

When all is ready to commence my footing course, I have mortar made, composed of one part cement to two of good sharp sand. I use this cement mortar until far above the ordinary water level, when I use lime one part, ashes one part, and sand three to five parts, just as I find the sand requires, for some sands are naturally hungry and require a large proportion of lime. No just rule can be laid down, only this: never let it get above five parts sand to one of lime. The ashes I use because they make the lime mortar set like a cement mortar and impart some considerable degree of hydraulic energy to it and render it impervious to the moisture, which would otherwise decompose the ordinary lime mortar. Either coal or wood ashes are good, and in their absence brick dust will be found no mean auxiliary to the lime.

When the work is completed I point up the joints with a mortar of one part cement and one part sand, into which is worked one part lime to make the mortar work easily and adhere where thrown.

CHAPTER V.

MANNER OF CONSTRUCTION—ABUTMENTS.

Having taken into consideration the height of embankment and pressure exerted outwardly upon any masonry erected to withstand this outward thrust, and having decided upon the shape and thickness of abutment, which thickness is also regulated to a great extent by the width of the foundations upon which we are to build, we turn our attention to the masonry. If the work is one of no great importance,—if the height of the opening, and consequently the weight of the masonry itself are slight, I care for nothing better than good solid rubble laid in ranges, roughly hammer-dressing the side joints into perpendicular and pointing or picking off any glaring inequalities on the beds; being careful to lay each stone so that it overlaps the joints of the work in the lower course, and at the same time spalling up any requiring it so as to put it upon a level with the others of the same course. Where the stones do not fit closely I fill the vacuities with broken stone and mortar, which in time become equally hard with the stone.

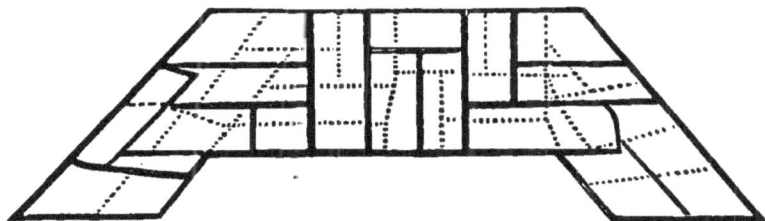

SECTION OF WING AND ABUTMENT.

The general construction of this work is shown in the illustration of footing courses of an abutment with diagonal wings, the

dotted lines showing about how the joints of the upper course
break those of the lower.

If the work is of still less importance, and the height of opening perhaps but four or five feet, consequently needing much
less strength, I should face my work with layer stone laid in
courses, with occasional headers laid through or at least nearly
through the wall, so as to bind it thoroughly. The back I
should build of the ordinary rubble or waste rock of my quarry,
and fill in between with the same or with a first class concrete of
broken stone. If good rich mortar is used in such work, the in-

ABUTMENT IN RUBBLE RANGE WORK—END ELEVATION.

terior of the mass will in course of time become much more solid
and harder than any sandstone. No trouble need ever be experienced from it,—I have used it as the filling of very heavy bridge
abutments and piers, the outside being faced with cut stone and
the headers running back and through, resting upon the concrete.
Then by the addition of more concrete the header becomes completely imbedded in it and so thoroughly cemented as to fracture
through the stone before parting from its concrete bed. For the
lower work cement concrete is needed; above that the lime concrete will answer. In making concretes I simply make up a bed
of mortar into which I shovel damp broken stone until the mor-

tar can take up no more. I hoe this mass over thoroughly so that each and every stone becomes well coated with mortar, and carry it into the work. This mass eventually becomes a conglomerate stone of great hardness, and where good stone is expensive can be employed to advantage, but usually it is more costly than solid stone work.

As before remarked, in some situations concrete is cheaper than well shaped stone, while in others it is costlier.

I vary my modes of construction for the very good reason that I wish my quarry kept clean and free from the waste stone and

'ABUTMENT IN RUBBLE RANGE WORK—VERTICAL SECTION.

dirt that are always accumulating. The earth which is uncovered from the stone I load upon the cars and set out upon some sidetrack until I have need of it—and I have invariably found occasions for the use of earth for filling around my work in finishing up—in which case the loaded cars come in very acceptably.

Again, between the layers of the rock are ledges of shale and shingle, which pick and blast out in very small fragments, and require so little exertion to break up into a size that will pass through a two or three inch ring, that I have had one man reduce an entire car load in one day. This as well as the junk stone I set out and use, because under such circumstances I can do it profitably and at the same time keep my quarry in working shape. It is not from preference that I do so, however, as the company has an excellent crusher located on the road from which I have never asked or received a single car of broken stone. That I prefer large stone is evident from the fact that I use the rubble and concrete only in the minor works.

Both classes of this work are shown on pages 30 and 31. All that is necessary to render them worthy of consideration is conscientious workmanship executed with first-class mortar. From motives of economy their adoption is rarely advisable unless under circumstances similar to those noted, and even then the constant supervision of the foreman is required to see that the work is well executed.

Upon most American roads masonry is a secondary matter and is generally built after the road has been opened. When first constructed piling and timber are freely used, but the road being put in operation then comes the substitution of masonry under all manner of drawbacks and obstacles. The stone must either be unloaded along the track at some distance from the work, upon a level grade, or put upon the side of some steep embankment at the work and also scattered in the dry bed of the stream. The dirt from the excavation becomes quite a hindrance, and the piles, bents or struts interfere greatly with the free working of the derrick.

The sand pile, mortar beds, water barrels, material, etc., all combine to make everything crowded and inconvenient. Sand, perhaps, must be brought from a distance; even the stream may be dry and water must be pushed on a car from some other stream or taken from the tank of some passing locomotive. Considerable allowance must therefore be made for such loss of time. Before beginning the excavation the track must be well protected from any danger which would ensue from the caving in of the embankment, which always happens more or less. This protection is afforded by putting in additional lengths of stringers which reach well back upon the embankment and which can be upheld near the excavation by bents or a crib of ties, so that the work can go steadily on without holding trains or delaying the work until they have passed over.

The mere building of the stone work generally occupies much less time than making ready for it and finishing up afterwards, and it is essential that such masonry should be built so as to be permanent, that all the confusion and expense may require no repetition. The entire matter is one needing the exercise of considerable judgment. There are cases where an excellent material foundation can be found at very slight depth, where there is no danger whatever from scour, and where the excavation of deep trenches would be a waste of money. There are others where the embankments need walling up with heavy wings or strong retaining walls. In some cases these wings should extend in height almost to the top of the grade; in others their height need be inconsiderable and the edge of the embankment should rest upon the tops of the wall, such being then a surcharged retaining wall.

Again, very often the sides of the embankments require no aid whatever in the form of stone work, or perhaps at most require some loose rubble thrown along their base. In fact, what will answer in one situation will not in another. In one an open culvert is preferable, in others a box or an arched culvert is better,—and these we will next consider.

3

CHAPTER VI.

BOX AND ARCH CULVERTS.

A box culvert is one provided with a flat roof or cover of stone, reaching from one wall or pier to another parallel wall and resting thereon. Such opening is of necessity narrow, and unless unusually long and heavy stones are used it can rarely exceed four to six feet in width, as it must be remembered that the covering stones must lap over upon the piers which sustain them. The embankment must also be of considerable height, for it is absolutely necessary that at least two feet of earth shall rest upon the roof between it and the ties. This earth acts as a cushion in softening the shock of a passing train. In the absence of a sufficient layer of earth the roof would be broken, not from the weight of the train but from the repeated hammering blows of the wheels.

The excavation for a box culvert should include not only the space to be occupied by the walls or piers but also all the intermediate space between. The bottom should be brought either to a uniform level or hollowed out in the centre from end to end and paved with good large stone that will not be washed out by the current.

The two sides of this pavement will serve as footing courses for the walls of the culvert. The covering should be of rectangular stones fitting closely together, so that in case the embankment is high each stone need bear only the weight of the earth directly above it. In general too little care is taken with this class of work. The excavation is made very slight, the culvert bottom often being considerably above the surface of the adjacent land, making drainage impossible but complaint from the land owner certain and earnest.

The roof is often made simply of several long stones laid a foot or more apart, like the rafters of a timber roof, and across from one stone rafter to the other is laid any rubbish in the shape of stone that will cover the space. As a consequence each long stone must support not only the weight of earth directly over it but also that over the stones that lap from it to its fellow. In

BOX CULVERT, WITH PAVEMENT.

the case of a heavy embankment such additional weight often proves too much and breaking in of the roof takes place, and in order to repair this damage an excavation perhaps twenty or thirty feet or more in depth must be made to get at the work and rebuild it.

Such work should be well built ot heavy stone, and every precaution should be taken against breaking in of walls or roof. I lately built a double culvert, one chamber of which was composed of only seven stones in its two walls and nine in the roof.

The extreme size and weight of these stones made them slow and tedious to handle, but the strength and permanence of the work amply compensated. Diagonal wings are essential to the entrance and also to the exit, and the space between should be paved, so that no holes can be worn by the rush of the water, which would

ARCHED CULVERT OF CUT STONE.

form eddies resulting in the undermining of the culvert proper and perhaps wearing a passage under the floor or along its sides, preparing the way for a washout.

A costlier and more important class of work is known as the arched culvert of which a sketch is shown below. In this, *a a* are the impost stones or tops of the piers from which the arch springs. The distance between these is called the span of the arch. The height from the level of the imposts up to the keystone *b* is called the rise of the arch, the inside curve is the soffit, while the outer or upper curve is called the extrados and each stone which is a member of the arch is designated a voussoir.

Before building an arch a timber framing or centering is put into position, and commencing at the imposts the voussoirs are laid upon this framing until all the keystones are inserted, which bind the work together so that the framing can be removed.

In this class of work accurately dressed stone should be used, not only in the arch itself but also in the piers. A reliable and uniform foundation is the only kind upon which such work should

ENGLISH BOND—EVERY ALTERNATE COURSE HEADERS.

be started, as any sinking of the haunches will tend to settle the lower courses of the arch and spring the voussoirs away from the keystones, weakening the arch and quite likely causing it to fall in. For this reason dressed stone should be used in the piers. Again, in the arch rubble stone could be used, but on account of the shape of the undressed stone much mortar will be required. This of course fills up all vacancies and irregularties, and acts as the key or wedge which holds the work together.

Now let this mortar decompose, soften, crack and fall out to any appreciable extent and the work must fall, because that which tightened it and held it together is gone, just as knocking out the wedges from a lot of blocking will loosen it.

The better plan is to cut the voussoirs into exact shape and rely upon this shape to uphold the arch and use only so much mortar as is necessary to cement the work together. On account of this accurate dressing, brick is now more frequently used than stone in the construction of arches, and where much of this class of work is to be done, instead of using the ordinary rectangular

HORIZONTAL SECTION THROUGH WALL.
(Showing headers running through the wall, and stretchers lengthwise with the wall.)

brick, beveled ones should be made, the arches being shaped to suit the bevel of the bricks—let the result be an eliptical, semi-eliptical, gothic or any other form of arch.

In former days many expensive stone bridges were constructed with arches of almost incredible span. The railroads of to-day find iron bridges resting upon stone piers preferable, and beyond culvert work, very few arched water ways are now built and these only where the height of the embankment makes them cheaper than open culverts. Therefore, little if any more need be said concerning them.

We will now turn to that class of work wherein not only all the difficulties of founding the piers of culverts are experienced, but where the presence of water acts as an additional obstacle—

where some mode of protection must be adopted against this troublesome element. The artificial foundations already illustrated can be used for the abutments and piers of heavy bridges, but when such work must be set in the bed of a stream, cofferdams, caissons, etc., are needed so that the foundations can be kept comparatively dry while the work progresses.

In the accompanying illustrations the manner of laying the stone is shown. The English bond is considered the strongest although either style will be found sufficiently massive if built upon a proper foundation. We are, therefore, left very little to consider but hydraulic foundations in future articles.

CHAPTER VII.

HYDRAULIC FOUNDATIONS

In carrying out the construction of bridge masonry over water courses of considerable size, in addition to the difficulty often experienced in obtaining good foundations, we encounter another great obstacle in the form of water. Various means must be adopted to make this drawback as inexpensive as possible. Every river bed may present some peculiarities of its own, or some combination, differing from all others, and it is these that must govern us in the modes of working. In one place we may use piling or sink caissons for foundations; in other places we may bring up solid masonry from the natural bottom, executing the work with the aid of the diving bell, or we may construct a cofferdam around the site of the work, and the interior being pumped dry, we can carry up the masonry above the level of the water line, when there is no further need of the dam. However, I will enter into no theoretical disquisition upon the matter, nor will I attempt to classify the various modes of founding and the circumstances which make them most desirable. I will, instead, explain my views much better by entering at once into actual practice, wherein the features of the river bed, banks and current are varied.

I wish to build the abutments and four piers for a bridge to span a certain wide stream as shown in the illustration. The site of the first abutment is back some fifteen feet from the edge of the water at its usual stage, in a heavy bank of stiff, shelly, blue clay which rests directly upon the bed rock some thirty feet below ordinary water level.

The first pier is located in ten feet of water, over-laying 20 feet in depth of mud and hard clay. and is somewhat apt to scour in time of freshets.

The second pier is to be built at a point where the bottom is perfectly smooth, level rock—with a depth of water of some thirty feet — the current being here extremely swift and often carrying considerable quantities of driftwood and ice.

At the third pier the level ledge of rock has terminated — the bottom is extremely rough, with broken bed rock, at a depth of fifteen feet.

There may also be a fourth pier (not shown in the illustration) rising from a small island or bar of mud and silt of semi fluid character, ten feet in depth.

The remaining abutment will be built in the

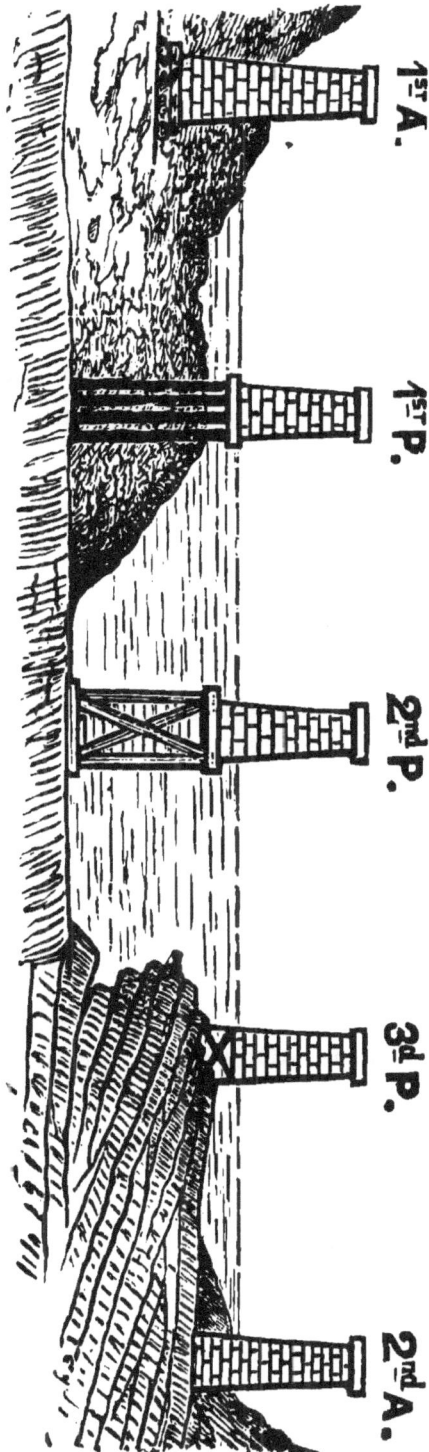

HYDRAULIC FOUNDATIONS—ABUTMENTS AND PIERS.

1ST A.   1ST P.   2nd P.   3d P.   2nd A.

water's edge upon hard level rock only three feet from the surface.

Commencing with the first abutment I excavate my pit to a depth of about twelve feet from the surface nearest the water, encountering no trouble from that element, except from the small quantities which seep in through several slight seams in the clay, on account of which I am compelled to pump out and sheath the entire work on one side with planking, behind which I ram down a casing of cement which cuts off almost all the water. At this depth I decide to commence the masonry, not thinking it necessary to carry the excavation down to the bed rock, because the pit is so far back from any danger of scour from the current that any further expenditure would be uncalled for. I would like to have set the masonry directly upon the clay or upon a light bed of cement concrete, but the leakage of water into the work has so softened the bottom of the pit that I make use of timbers laid transversely with the trench and filled in between with concrete and planked over, as has already been illustrated in one of the chapters on culvert work. As each course of masonry is laid I pack the space between it and the sides of the pit with well puddled clay, which I see is well rammed down. The masonry is then carried up in cement mortar nearly or quite to the finish in the mode already explained.

This is not therefore properly a hydraulic foundation, but in the execution of the first pier I find my work must be carried up through twenty feet of stiff clay and mud, over which flows at ordinary stages some ten feet of water. At considerable expense I could build a cofferdam around the site of this pier, and either start the masonry from the bed rock or set it down in the clay, as with the first abutment, but I do not desire to go down to the bed rock on account of the cost; and again, on the other hand, I notice some little evidences of scouring which create doubt in my mind as to the permanency of any work merely set down into the clay, as it would evidently stand in danger of being undermined sometime. Therefore as the clay and mud are sufficiently dense to hold up in the perpendicular any piles that I might drive, I put in a pile foundation such as already shown in a previous chapter on culverts.

Bed rock being within easy reach of a thirty foot pile, instead of using short piles, which would penetrate only a part of the depth of the clay, I make use of those which will strike the rock and act in reality as posts, in which case, provided the piles are of sufficient strength, no settlement in the pier would be visible except that caused by slight shrinkage of the mortar in the horizontal joints of the masonry. But were the bed rock at much greater depth I would not drive the piles with the intention of resting them as in this case upon the solid rock.

Elaborate tables have been prepared by eminent authorities by which we are supposed to determine the safe load a pile can carry —as for example, "the force of the blow by a ram of certain weight falling a certain distance being given, to find the dead weight or pressure which the pile would sustain, etc." But in actual practice these formulæ and rules often prove sadly deficient in accuracy, for in the majority of these tables the calculations assume the soil to be homogeneous or of the same nature from top to bottom, whereas in reality there often occur cases where the piles pass through strata of alluvium, shale, clay, sand, hard pan, etc., and although they may resist the hammer, it is sometimes difficult to determine how much of this resistance is due to mere lateral friction, and how much to a hard stratum.

The better way in starting such work is to drive a test pile or so in order to determine, if possible, the required length for the regular foundation. Generally a pile should be driven until it will not sink more than one inch under the last blow of the monkey. Piles have been driven only twenty feet in stiff clay and sustained a weight of eighty tons, but not more than one-quarter that weight is often required to be upheld by any pile, and in mud and marsh bottoms care must be taken to drive it sufficiently deep so that the lateral friction alone can be relied upon to offer considerable resistance to the sinking of the pile under a reasonable amount of dead weight. Foremen of the pile drivers, if experienced, generally know best about the soils and bottoms along the line of the road which employs them. They have generally some pretty clear ideas about what manner of deposit the pile is penetrating—know when the pile has set and are quite apt to be very positive when they have struck solid rock, and their judgment can generally be deferred to with a conviction that they are not very far out of the way, if at all. In passing through some strata it is often necessary to ring the heads of the piles with iron bands or rings, to prevent splitting the piles under the blows of the hammer, also to shoe the points with cast or wrought iron, sawing off the extreme point of the pile, and making a shoulder or socket in the iron shoe into which this sawed off end will fit. Such a shoe will stand considerable heavy driving though much resistance be offered. If, however, the pile is sharpened down to a point and the iron fits close to it the sharp point will act as a wedge and split the shoe.

But I will return to the pier. When all the piles are driven they are cut off to a uniform level several inches below the low water line, the spaces between perhaps left open to the water or closed up by sheet piles, between those piles forming the sides of the work, and the interior vacancies filled up with broken stone, or under some circumstances, with cement concrete; timbers are then laid upon the tops of these piles and planked over as a platform on which to lay the masonry. The piles must be cut off sufficiently low to place all the woodwork permanently under water in order to prevent decay. If a cement concrete filling is used considerable additional expense must be incurred in order to prevent the washing out of the cement before it has set sufficiently to withstand the water, therefore, I rarely use it in such situations but adopt another method, as will be shown in piers 2 and 3.

The second pier of this bridge must be founded upon a smooth level bed of rock, lying at the depth of thirty feet below the usual stage of the water line. The surface of this rock is water worn and smooth from the action of a powerful current which prevents any lodgment whatever of sediment, mud or sand. The illustration already given shows the site of this pier in the bottom of a ravine or wide fissure in the river bed, forming its channel, through which is carried much driftwood, not only upon the surface but also partly or entirely submerged. A large body of such float striking and pressing against a pier merely setting upon this smooth level bed rock without fastening of any kind, would be very apt to slide it out of position, and in our work we must take some precautions against this danger. It is desired to build the superstructure with spans of uniform length, otherwise the pier might be located back farther in a less exposed position.

Piling is entirely out of the question because of there being no soil into which piles could be driven. I find the current so strong that a man in diving armor can do nothing unless its force is diverted. The only feasible plan is to construct a caisson or huge chest either with or without bottom. The natural surface of the rock being smooth and level, the ex-

pense of a timber bottom to the caisson is unnecessary.   In fact
these bottoms are never advisable or needful except in two situations: where the caisson is to be set upon piling, or where
we are building upon marshy or peaty ground of great depth, into
which we expect the caisson with the masonry inside to sink until
it compresses the soil under it so as to suspend the caisson in its
midst, or allow it to settle down uniformly until it eventually
touches hard pan or some other firm stratum, of which I shall
fully treat hereafter.

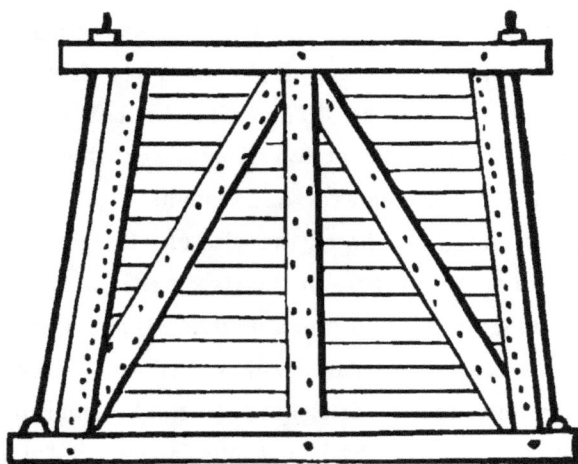

END SECTION OF CAISSON.

A caisson with timber bottom must rest upon a uniformly
level surface, for if it is deposited upon a firm stratum of uneven
surface, some parts of the bottom timbers will be unsupported
and experience severe cross strain, which will in time surely lead
to very serious fractures in the masonry.   Then again, if placed
upon any soft soil that becomes exposed to the current, the work
will be undermined and dangerous, and breaks will occur.   If the
bed be smooth, level and hard, we do not need the timbered bottom, because we can by a little scribing of the lower edges of the
sides and ends of a bottomless caisson place it so well in position

that with the aid of cement concrete we can make it almost perfectly water-tight. Therefore on the shore we construct the sides and ends of the caisson in panels or sections so as to be detachable after our pier shall have been completed. From two to four inch plank will suffice, according to the force and depth of the stream in which it is to be placed. The cross and upright timbers should be from four to six inches in thickness. These sides and ends can be erected upon ways which slope from the shore to some distance into the water at such depth as will float the caisson off the ways, when it can be towed out to its location and sunk by swinging rails on to the sides near the lower edges, these rails being drawn up again when the caisson is no longer needed.

The trouble of properly locating such a caisson is not so great as may be imagined. Any body under water can be moved with much greater ease than when above ground. While sinking, it can be held well in position by ropes made fast at the corners and can be shifted by them or moved around on the bottom by men in diving helmets.

Our caisson is put up in one piece, well caulked with tow and located upon the bottom. Our next move is to form a water tight floor upon the bed rock inside the caisson walls to stop off the bottom water and allow us to pump out the caisson and begin the masonry. In order to accomplish this, we deposit a layer of cement concrete only one foot in depth. Should we bring our flat boat alongside the caisson and shovel the concrete into the caisson—before reaching the bottom—in passing through the water, all the cement would be washed out and separated from the sand and broken stone forming the concrete, thereby rendering the concrete worthless. We therefore lower it by means of a box holding one cubic half yard, this box having a hinged bottom

whereby we can deposit the concrete immediately upon the bed
rock and in the water so near the bottom as to cause no percep-
tible disintegration.　The box being emptied quickly rises
to the surface to be returned as before.　We make use of
strong concrete of one part cement to one part sand, with
such amount of broken stone as can well be incorporated with it.
The edges of the caisson fitting closely to the rock, there is in
consequence but little scouring out of the concrete.　Not wishing
to injure it by the disturbance of the pumps we let it rest and in
three days time it is thoroughly set and the caisson is pumped
dry—when we find the layer of concrete does not present a uniform surface and is set so hard that we are compelled to point it
off before commencing the laying of the stone upon it.

The current seems to exert all its force to shift the caisson,
weighted as it is with the iron rails, and we begin to realize the
danger in which our pier would be placed in time of freshet.
Therefore we drill through the layer of cement eighteen inches
into the bed-rock, drilling four rows lengthwise with the base of
the pier, and at distances of four feet apart in the rows.　In
these drill holes iron dowell pins four feet in length are inserted
and leaded, the ends above the concrete layer which enter drill
holes in the stones of the footing course of the masonry being also
leaded.　The cut shows the footings of the pier and the shape of
the work with projecting base resting upon the footings.　From
this base the pier is built up with a cutwater of smooth, dressed
stone, well bonded into the heavy portion of the pier, being
gradually drawn in until it is merged into the pier at five feet
above high water line when both ends of the pier are brought to
a point in the form of circular arcs, having a radius of about
three-fourths the width of the pier.

In pier No. 3 the bottom lies upon the cropping out of

several ledges of rock, very rough and broken. We here sink
a bottomless caisson, whose lower edges we must scribe to a considerable extent so as to fit to some degree the irregularities in
their surface. A diver must be sent down to work off
some of the most prominent projections and fill up the fissures with rubble stone: but in spite of all this we cannot fit the
sides of our caisson closely to the rock. We therefore line it
with canvas with a rather baggy bottom, sink it into position,
and deposit our concrete within the sack, which adapts itself suf-

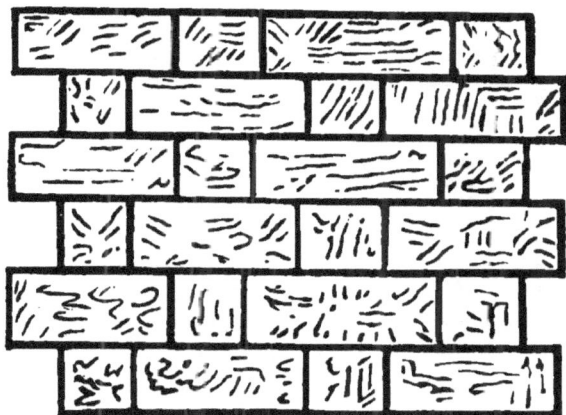

FLEMISH BOND—HEADERS AND STRETCHERS ALTERNATELY, IN
EVERY COURSE.

ficiently well to the surface: then we proceed as related in the
case of the second pier. This canvas bottom is a protection
for the concrete, as it would otherwise be washed away by the
current which would enter between the caisson and the rocks,
and would scour out nearly if not quite all of the concrete.

If there is to be a fourth pier (not shown in the cut) on a
bar of silt ten feet in depth, we again use the bottomless caisson,
which we sink with rails, dredge out and put in a canvas bottom.
Were this island composed of strong earth that would sustain
sheet piling, we would erect a timbered cofferdam, or were the

bed of silt only one-half so deep. we could surround the site with a cofferdam of clay. The second abutment of our work is to be located in the water's edge with a depth of only three feet to bed-rock. A clay bank being right at the work, our cofferdam is constructed of this. Two sluice boxes are put in so that the enclosure can be flooded with water at any time that a freshet is apprehended, and prevent any breach in the dam.

A cofferdam is an enclosure built around the site of the work for the purpose of pumping it out and keeping it dry

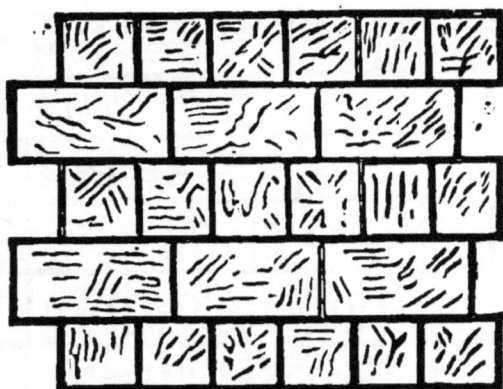

ENGLISH BOND—ONE COURSE ENTIRELY OF HEADERS; NEXT ENTIRELY OF STRETCHERS.

while masonry is under way. It is necessary that these enclosures be as nearly water-tight as possible. They are only adapted to depths of less than twenty feet, and the material of which they are constructed must depend not only upon the height of the dam required to keep out the water, but also on the character of the bottom. There are situations, as in case of our pier. No. 4, where cofferdams are not advisable, although the depth of excavation is not great. Very high cofferdams of earth are expensive, considerably more so than timber dams, but in slight depths they are often the best, cheapest and most expeditiously constructed. In still water, less than seven or eight feet deep, tarred canvas well weighted at the bottom will answer every purpose if stretched tightly and secured to piling.

But the cofferdam most generally built is one of either a single or double row of piling, the first piles being the guide piles driven in a row at intervals of about ten feet. If the depth is not great nor the current heavy, we may use only one row of piles; therefore, on each side of these guide piles, at the top and as near the bottom of the excavation as possible (see illustration), we bolt timber strips or wales, horizontally, and down between these wales we drive sheet piling, each joint as close to the other

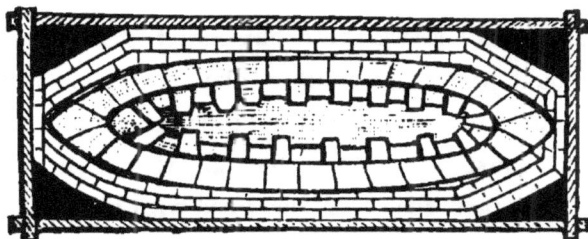

BASE OF PIER, AS BUILT WITHIN THE CAISSON.

as it can be driven, so that but little caulking need be done to make the joints water tight. Two-inch plank is sufficiently heavy sheet piling for some situations; in other instances square timber of six inches up to twelve in thickness may be required, according to the depth and pressure of the water. But in works of any consequence a double row of such piling is better, set with a width of at least one foot vacant space between, which we fill and ram down as it is being filled, with a well-worked mixture of wet clay and gravel, which is better and forms a more impermeable filling than clay alone. The depth to which these piles must be driven is only so much below the bottom of the work as will give them power to stand perpendicularly until the work is braced from the top—and also to prevent them from being pressed in at the foot by the head of water and soil. I have seen cases where a driving home of only three feet proved an excellent dam, which withstood a head of water of seven feet, although I wish it remembered that the tops of piles were braced from the inside.

But these piling dams cannot always be used, it being as impossible to obtain a foothold for them in soft mud, as it would be on rock bottom. So there are then two situations where piling is of no advantage. If the depth is too great for a clay dam, the only other resort is to make the dam of narrow bottomless caissons, sink them in a line around the site of the work and fill them up with puddled clay and gravel; but rather than do this most builders would prefer some other plan of work. Cofferdams must be adapted in strength to the head of water and soil whose pressure they must withstand. Generally they must be provided with sluices so that they can be filled with water on the approach of high tides or freshets, as it is better to pump the cofferdam out than to have it washed away or greatly damaged.

Cofferdams can be built upon rock bottoms at slight depths by using iron rods instead of piles. These rods are set into drill holes in rows and planking is laid between them. That they are expensive and not likely to come into general use can be easily seen.

At great depths cofferdams cannot be used; neither can the ordinary forms of caissons be used in excavations through extensive sand bars or great beds of mud. There are many rivers, for instance, the Arkansas, which are choked up with very deep bars of sand, which has been from time immemorial washing down from the great plains and filling up the original channel of the river, in many places to a depth of forty to sixty feet; indeed I think no bridge could be built over this stream south of Great Bend, the excavation of whose piers could be less than thirty feet in depth through coarse white sand, so loose that a dredging boat could work an entire year and still not a single cubic yard of excavation be credited to its account, unless the water was at a very low stage.

The Platte is a similar river, all of whose bridge piers have been put in by the pneumatic process, at great expense, and in

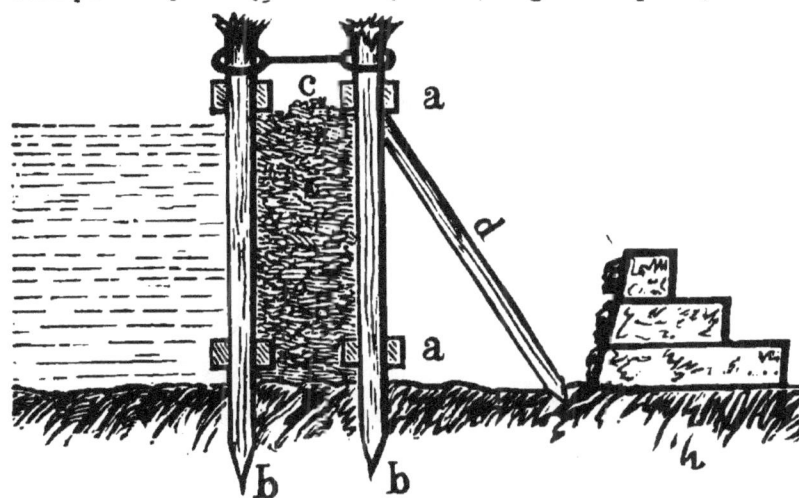

PILING COFFER-DAM.
Two rows of sheet piles (b) driven between horizontal wales (a) which are bolted to guide piles; c is the clay puddle or filling; d is the brace on inside of work.

some instances with considerable loss of life. The Missouri and Mississippi are rivers whose deposit is not sand but mud. Upon them also the compressed air process is used, attended by great loss of life. On the St. Louis bridge, under a pressure of three atmospheres, several men were paralyzed or died, and the working hours per day were gradually reduced from four hours to one. Of this system I know nothing from experience. I cannot treat of it therefore, but will illustrate a sinking caisson which I have used, and which can be gradually sunk to any depth and which is eminently adapted to work in great depths of mud or sand. The only obstacle to its perfect working is the encountering of a large snag or boulder, but these do not often occur in actual practice. If any doubt is entertained, several borings can be made over the

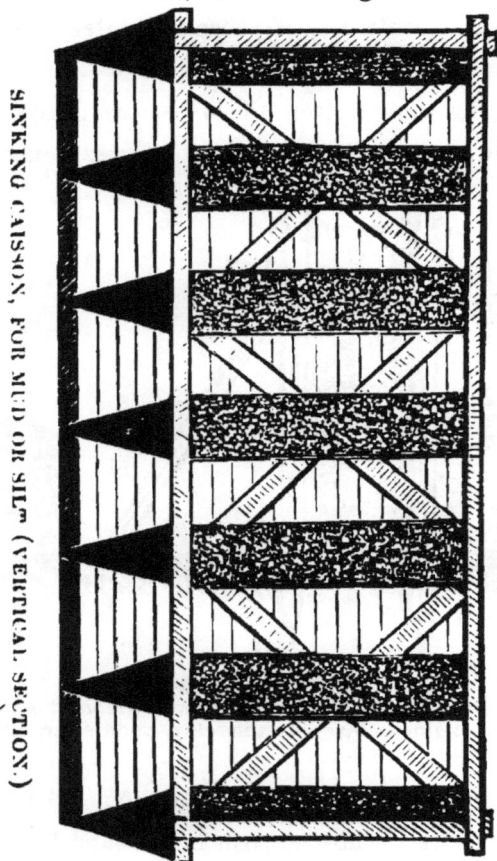

SINKING CAISSON, FOR MUD OR SILT (VERTICAL SECTION.)

site of the work. Any slight obstruction need not deter us in the use of this caisson, as the dredging chambers afford considerable scope for working away any small snag or boulder.

This caisson is made in two portions, the bottom having of course four sides, some five to ten feet high, built of timbers in such shape as to make the top portion of each side some three to ten feet wide, from which it tapers down to a cutting edge at the bottom, which is shod with iron. These four sides are cut up by the addition of other walls built exactly the same. In the cut I divide the caisson lengthwise into two equal parts, then transversely I divide it by five partitions of similar construction into altogether twelve compartments. The bottom of the caisson is now ready for building upon it the upper portion, the outer sides being constructed as in other forms of caissons, with the exception that parallel to these sides, upon the outer edge of the head of the wedge shaped bottom, is built a stout plank lining extending up to the top of the caisson, so as to form a hollow wall around all four of the sides. Up from the heads of all the cross partition wedges are built hollow walls with perpendicular sides. This, therefore, gives us three long compartments—two at the ends, and ten cross sections, whose bottoms are the timber heads of the wedge shaped sides and partitions. In the intervening space of the interior of the caisson we have twelve other chambers which are entirely open at the bottom and whose sides are perpendicular in the upper portion. After that, in the lower portion, the sides flare open towards the bottom or mouth, this flaring shape being given by the wedge shaped sides and cross sections.

When all is ready the caisson is located over its site and the compartments within the hollow walls are filled with concrete, timbers being laid in lengthwise with each wall every six feet, so that those in the transverse sections will cross over those in the lateral walls and bind the various columns of concrete together. As the concrete filling takes place the edges of the caisson bottom cut into the soil and the work continues to sink until some little time after all the closed chambers are full of concrete. The caisson can now be weighted with rails and sunk still deeper, or we can at once commence dredging

through the chambers which had no bottoms and which were de-
signed as dredging shafts. As the caisson sank from the weight
of the concrete, the wedge shaped sides to the mouth of each
chamber pressed the soil in towards its center and on up the perpendicular shaft. A caisson, say forty feet only in length, with
fifteen chambers of concrete, each chamber being five feet in
width, must cleave the mud or sand through which it passes to
considerable extent, when we reflect that the twelve dredging

SINKING CAISSON.
Lower section, showing dredging chambers and concrete compartments.

chambers not nearly so large as the concrete compartments receive into their funnel shaped mouths and for some distance up
each dredging shaft all the soil that was under the entire caisson.
This it does and brings it into easy reach of the dredger. In
such a caisson the builder can proportion it to suit himself. He
must have sufficient storage room for such amount of concrete
as will weight the caisson, so that it will sink as the dredging
goes on. He must allow in the dredging chambers ample scope
for the dredger to operate. He must weight the caisson with
the concrete so that it sinks evenly, and in the dredging take the
same care and try to sink the work as uniformly as possible.
When the bottom is reached and the chambers all dredged out
they should be filled with concrete to the top, level in all the
compartments, when the masonry should be started thereon.

CHAPTER VIII.

CONCRETE WORK.

Along the lines of some railways, procuring stone suitable for the general purposes of railway masonry is attended with so much difficulty as often to render its use practically impossible,—but sand, gravel or rough stone is more or less abundant in every locality—and either of these or all three, combined with the proper proportions of hydraulic cement and water, well incorporated with each other, can be used as a substitute for stone work and can be run into timber moulds or frame work and adapted to all forms of masonry construction with success.

The excellence of concrete depends greatly upon the materials which are used in its composition in combination with the cement. Sand can be used alone, but as the proportion of this should not exceed three parts to one of cement, the bulk of the concrete would be so small that it would be found to be very expensive, especially when fully as good, if not better conglomerate can be made by the addition of at least five parts of gravel or broken stone; in fact, all that is required is a sufficient quantity of the bonding material (that is, the cement mortar) to unite the various particles of stone or gravel. The closer these particles can be made to adjust themselves the less bulk need be occupied by the bonding substance in order to form a solid mass. Small angular stones if pressed together will adjust themselves more compactly than ordinary rounded gravel stones; but when loosely dumped in a heap the interstices among the angular stones are much larger than in the gravel—whereas the gravel—such as is often used in roofing—when tipped into a heap will settle itself

into as great compactness as it is capable of. The ramming
of clean gravel, instead of being beneficial, tends to displace the
various members without bringing them into more compact relation.   If we use for bulk neither gravel nor broken stone but
shingle stone, which, as its name implies, is in flat, thin pieces,
and being neither angular nor rounded, can be packed with considerable advantage as the mass of concrete is being put into the
moulds. Concrete walls in buildings possess one great advantage
over either brick or stone in being more porous than either of these
and a very poor conductor of heat or cold.   They show no frost
on the inside in the winter and are drier and cooler in the sum-
mer.   In ordinary building the concrete walls should be of somewhat less thickness than brick walls, and far less than stone.   I

CONCRETE ABUTMENT, FACED WITH CUT STONE.

here introduce an illustration of an abutment proper, not a mere
shore pier, but a work that actually abuts upon the embankment
and not only upholds the superstructure, which a shore pier does,
but also acts as a retaining wall in supporting the embankment
head, which a shore pier does not do.   This work is faced with
cut stone, although concrete, moulded into blocks would answer
equally well.   The courses of cut stone were laid up only one or
two in advance of the concrete backing,—this stone facing there-

fore acting as a mould for the concrete—rendering timber shutters or frames unnecessary. The excavation for the foundation was made with a level floor, into which huge saw-teeth cuttings were made—so as to more fully insure the stability of the abutment and its power to withstand the heavy outward pressure of the embankment,—for while concrete is heavier than earth and less liable to slip when used in retaining walls or embankments, yet in no case will its weight equal the same bulk of stone similar to that of which it is partly composed. Concrete varies in weight according to the density of the stone which enters into it, 120 lbs. per cubic foot being about a fair average, while 96 lbs. is a large allowance for earth. Its value as a material for retaining earth embankment is therefore readily seen. But when we compare it with basalt weighing 180 lbs., or granite weighing 165 lbs. to the cubic foot, ample reason can be seen for the increased bulk which concrete must occupy in order to resist as much pressure as the same weight in solid stonework. The precautions shown by the shape of the work are therefore explained.

I have already dwelt upon the making of concrete and the care that must be taken not to separate the cementing substance from the other ingredients by dumping from any splashing height or through water. There may be some occasions where honeycomb concrete work can be used, i.e., where larger sized broken stone is used and only enough mortar mixed in to coat the various stones over and cause them to adhere whenever they come in contact, leaving many interstices between; but generally solid work is preferable in building these abutments. Instead of filling up the mould at one time, and as quickly as possible, I put it up in successive layers of equal height with the courses of cut stone in the facing, my object in depositing the concrete in layers being that in case the bottom upon which it is laid should settle,

only such layers as were already in the work would settle with
it. Any extensive settling would result in cracking the concrete
then in the work. If the entire mass were in position at the
time of such settling, a fissure or fissures would be noticeable extending from the bottom to the finish, or nearly so, being of
course greater in width at the bottom. This would probably separate the concrete into several distinct parts utterly without
bond, thus weakening the work. But some one may remark that
the same settling might occur in the other mode, and that it
would not be apt to take place until considerable weight of concrete had been used; all of which is true. But should the settling occur when nearly all the layers had been run in, at least
the upper portion of the fissures could be filled with thin mortar, called grout, and the next layer of concrete would adapt itself to the depression of the main bulk. Again, in layer work
these fissures are not perpendicular, but zig-zag, somewhat similar to the same in stone work, the bond being still preserved
but lacking cohesion. In heavy work, which depends greatly
upon its weight as a power of resistance, no great injury ensues,
but in a light wall depending more upon remaining perpendicular, the impairment of its strength by such cracks often results
disastrously.

No matter how carefully foundations are prepared, something
of this kind may occur now and then; the bottom may not present a uniform degree of compressibility. Especially is this noticeable in repair work. For instance, old work being torn out
and enlarged work put in, the extension, resting upon a new bottom, is more liable to settle than is that portion formerly weighted
and compressed by the old work. In building with concrete any
carpenter would easily understand the mode of putting up stand-
ards in which the plank framing can be inserted and from which

it can be removed. The bed for concrete may be upon timber, or in case of a floor, well rolled gravel is proper. although in any pavement or floor great pains must be taken in preparing the bottom, or the work will certainly crack.

I will not undertake to explain this subject farther. The principles remain the same, whether a tank, cistern, abutment or ordinary wall is to be constructed.

CHAPTER IX.

PROTECTION AGAINST SCOUR.

The banks and beds of some streams are composed of soils very susceptible to scour by the current. Especially is this the case where the material is loamy earth or sand, these beds being often of such depth that there seems almost no limit to their washing out. Owing to the eddies and whirlpools created by

PIER PROTECTED BY CRIB WORK.

such obstructions as piers, the greatest scour can be anticipated directly at and around the base of the pier. The only inexpensive method of protection in such a case is to deposit a heavy layer of large boulders all around the pier. In most instances this will suffice, but should the current be very strong, these loose boulders will quite probably be washed down the stream again, leaving the masonry exposed. The only remedy then is to crib up a certain space around the pier with heavy logs and fill it with boulders or a very heavy bed of concrete. The latter plan is the most effective of all, but an apron of concrete three or four feet in depth is a very expensive affair, equalling in cost a like superficial area of the bridge above it. Unless absolutely required to clear the channel, piles which have been driven around the site of a pier should never be drawn. It is better by far to allow them to remain and fill up any vacant space with rip-rap, the drawing of a pile leaving a cavity as dangerous as any exca-

PIER WITH APRON OF HAND LAID RIP-RAP.

vation. In some cases the pier can be protected from scour by driving piles around it and filling the space with rip-rap. If no large stones are available, a bed of broken stone can be thrown around the work and this paved over with hand-laid rip-rap of stones about eighteen inches or more in depth when placed upon their edges, as per illustration. This manner of laying rip-rap is especially adapted to the banks of a stream.

If well under water all this work must be done by men in diving armor at very high wages and with several attendants to keep the air pumps in operation and to await the signals of the divers. The manufacturers of diving apparatus charge extraordinary prices for very little, the main part of which can be gotten up in any company's shops at very small expense comparatively.

I hardly think it advisable to give any directions or illustrations concerning the making of diving outfits. I unfortunately got up an experimental suit once upon a time and as a result had to make constant use of it for twelve days, and found it anything but agreeable for an amateur. It is one of those cases where "'tis folly to be wise," and where the masons can step aside and allow the professional diver full scope for the exercise of his ability. However, should any of my readers feel very much disposed to do a little deep water work, I will cheerfully furnish them all necessary information free of charge and wish them all success in their undertaking.

In concluding these articles on bridge foundations and masonry I will only mention the necessity for carefully judging the proper mode of founding, which must depend upon the nature and depth of the bed of the stream. I have endeavored to outline the principal methods of providing foundations and protecting them, but in the rush of active work I have found but little time to enter into details as freely as I could wish. Experience is the great teacher, and it will furnish all the minor information which I have overlooked, and also much which I have yet to learn.

www.ingramcontent.com/pod-product-compliance
Lightning Source LLC
Chambersburg PA
CBHW021524090426
42739CB00007B/769